Florian Ion **PETRESCU**

Some New Elements in Physics

USA   2011

Scientific reviewer:

Prof. Dr. Eng. Nicolae Mihăilescu

Copyright

Title book: Some New Elements in Physics

Author book: Florian Ion PETRESCU

© 2011, Florian Ion PETRESCU

petrescuflorian@yahoo.com

ISBN 978-1-4679-4880-7

Welcome! A Short Book Description

The movement of an electron around the atomic nucleus has today a great importance in many engineering fields. Electronics, aeronautics, micro and nanotechnology, electrical engineering, optics, lasers, nuclear power, computing, equipment and automation, telecommunications, genetic engineering, bioengineering, special processing, modern welding, robotics, energy and electromagnetic wave field is today only a few of the many applications of electronic engineering. This first chapter presents shortly a new and original relation which calculates the radius with that the electron is running around the atomic nucleus.

The second chapter presents, shortly, a new and original relation (20) which calculates the Doppler Effect exactly. This new relation (20) is the exact form and the classical expression (10) is an approximate relation.

Renewable energy is energy which comes from natural resources such as sunlight, wind, rain, tides, and geothermal heat, which are renewable (naturally replenished). The share of renewables in electricity generation is around 18%, with 15% of global electricity coming from hydroelectricity and 3% from new renewables. The third chapter aims to disseminate new methods of obtaining energy. After 1950, began to appear nuclear fission plants. The fission energy was a necessary evil. In this mode it stretched the oil life, avoiding an energy crisis.

3

Even so, the energy obtained from oil represents about 66% of all energy used. At this rate of use of oil, it will be consumed in about 40 years. Today, the production of energy obtained by nuclear fusion is not yet perfect prepared. But time passes quickly. We must rush to implement of the additional sources of energy already known, but and find new energy sources. In these circumstances this chapter comes to proposing possible new energy sources, like energies obtained by the annihilation of a particle with its antiparticle.

PRESENTATION

CHAPTER I - PRESENTING OF AN ATOMIC MODEL AND SOME POSSIBLE APPLICATIONS IN LASER FIELD

The movement of an electron around the atomic nucleus has today a great importance in many engineering fields.

Electronics, aeronautics, micro and nanotechnology, electrical engineering, optics, lasers, nuclear power, computing, equipment and automation, telecommunications, genetic engineering, bioengineering, special processing, modern welding, robotics, energy and electromagnetic wave field is today only a few of the many applications of electronic engineering.

This first chapter presents shortly a new and original relation which calculates the radius with that the electron is running around the atomic nucleus.

CHAPTER II – SOME FEW SPECIFICATIONS ABOUT THE DOPPLER EFFECT TO THE ELECTROMAGNETIC WAVES

This chapter presents, shortly, a new and original relation (20) which calculates the Doppler Effect exactly. This new relation (20) is the exact form and the classical expression (10) is an approximate relation.

The classical approximate relation (10) presented in the form (15) can't foresee the Doppler Effect for the case when the angle $\varphi=90^0$. For this reason it was introduced the relativity effect, where the period T_0 take the form T_0/α. Before to utilize the theory of the relativity it's strongly necessary to test the relations (8), (18) or (20), and the particular form (14) (for the angle $\varphi=90^0$), to testing the Doppler exact effect without the relativity theory.

The Doppler Effect represents the frequency variation of the waves, received by an observer which is drawing (coming), respectively it's removing (going), from a wave spring (source).

If a bright spring is drawing to an observer, the frequency of waves received by the observer is bigger than the emitted frequency of source, such that the respective spectral lines are moving to violet. On the contrary, if the light source is removing from the observer, the spectral lines are moving to red.

One proposes to study the Doppler Effect for the light waves, generally for the electromagnetic waves. The paper proposes for the Doppler Effect the relation (20) which can replace the classical form (10).

CHAPTER III - The Future Energy

Renewable energy is energy which comes from natural resources such as sunlight, wind, rain, tides, and geothermal heat, which are renewable (naturally replenished). In 2008, about 19% of global final energy consumption came from renewables, with 13% coming from traditional biomass, which is mainly used for heating, and 3.2% from hydroelectricity. New renewables (small hydro, modern biomass, wind, solar, geothermal, and biofuels) accounted for another 2.7% and are growing very rapidly.

The share of renewables in electricity generation is around 18%, with 15% of global electricity coming from hydroelectricity and 3% from new renewables. This chapter aims to disseminate new methods of obtaining energy. After 1950, began to appear nuclear fission plants.

The fission energy was a necessary evil. In this mode it stretched the oil life, avoiding an energy crisis. Even so, the energy obtained from oil represents about 66% of all energy used. At this rate of use of oil, it will be consumed in about 40 years. Today, the production of energy obtained by nuclear fusion is not yet perfect prepared.

But time passes quickly. We must rush to implement of the additional sources of energy already known, but and find new energy sources. In these circumstances this chapter comes to proposing possible new energy sources, like energies obtained by the annihilation of a particle with its antiparticle.

CHAPTER IV - NEW AIRCRAFT

Speaking about a new ionic engine means to speak about a new aircraft. This chapter presents shortly the actual ionic engines (called ion thrusters) and the new ionic (pulse) engines proposed by the author.

Ionic engine (ion thruster, which accelerates the positive ions through a potential difference) is about 10 times more effective than classic system based on combustion.

We can still improve the efficiency of 10-50 times if one uses pulses of positive ions accelerated in a cyclotron mounted on the ship; the efficiency can easily grow for 1000 times if the positive ions will be accelerated in a high energy synchrotron, synchrocyclotron or isochronous cyclotron (1-100 GeV). In this, the big classic synchrotron is reduced to a ring surface (magnetic core). Future (ionic) engine will have mandatory a circular particle accelerator (high or very high energy).

We can thus increase the speed and autonomy of the ship using a less quantity of fuel and power.

One can use synchrotron radiation (synchrotron light, high intensity beams), like high intensity (X-ray or Gamma ray) radiation, as well. In this case will be a beam engine (not an ionic engine), it'll use only the power (energy, which can be solar energy, nuclear energy, or both) and so we will remove the fuel. It proposes using a powerful LINAC at the exit of synchrotron (especially when one accelerates electrons) to not lose energy by photons premature emission.

With a new ionic engine one builds a new aircraft, which can travel through water and. This new aircraft will can accelerate directly, without an additional combustion engine and without gravity assists from other planets.

CHAPTER V - CAPTURING ENERGY CONCENTRATED NEAR THE SOURCE AND FORWARDING DIRECTLY TO EARTH IN CONCENTRATED FORM

Should start some spatial projects, to capture a large amount of energy somewhere near the source (near the Sun), energy which can be sent then to the Earth in a concentrated form (LASER, MASER, IRASER, etc).

The enormous energy emanating from the sun is spreading in all directions of the universe, and dilute with the distance.

On Earth no longer reach than a small amount from the energy emanated by the sun.

We try here (on the Earth) to capture a drop from a very small amount of energy, who came from Sun. And we also complain that the yield is low, and technological costs are high.

Installations which must do capturing the solar energy, could be installed over the Mercury.

From the Mercury, the concentrated energy will be transmitted directly focused on the Moon.

On the Moon, the energy will be conserved and forwarded to Earth in doses non-hazardous (with lower concentrations), using multi-channels microwaves.

CHAPTER I - PRESENTING OF AN ATOMIC MODEL AND SOME POSSIBLE APPLICATIONS IN LASER FIELD

INTRODUCTION

This chapter presents, shortly, a new and original relation (20 & 20') who determines the radius with that, the electron is running around the nucleus of an atom [2].

In the picture number 1 one presents some electrons that are moving around the nucleus of an atom [1].

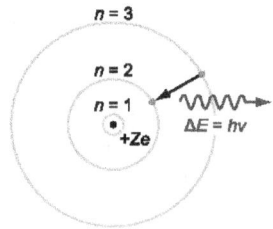

Fig. 1 *Electrons moving around the atomic nucleus;*

The atomic nucleus consists of nucleons (protons and neutrons)

One utilizes, two times the Lorenz relation (5), the Niels Bohr generalized equation (7), and a mass relation (4) which it was deduced from the kinematics energy relation written in two modes: classical (1) and coulombian (2). Equalizing the mass relation (4) with Lorenz relation (5) one obtains the form (6) which is a relation between the squared electron speed (v^2) and the radius (r).

The second relation (8), between v^2 and r, it was obtained by equalizing the mass of Bohr equation (7) and the mass of Lorenz relation (5).

In the system (8) – (6) eliminating the squared electron speed (v^2), it determines the radius r, with that the electron is moving around the atomic nucleus; see the relation (20).

For a Bohr energetically level (n=a constant value), one determines now two energetically below levels, which form an electronic layer.

The author realizes by this a new atomic model, or a new quantum theory, which explains the existence of electron-clouds without spin [1-2].

Writing the kinematics energy relation in two modes, classical (1) and coulombian (2) one determines the relation (3).

From the relation (3), determining explicit the mass of the electron, it obtains the form (4) [2].

$$E_C = \frac{1}{2} m \cdot v^2 \qquad (1)$$

11

$$E_C = \frac{1}{8} \frac{Z \cdot e^2}{\pi \cdot \varepsilon_0 \cdot r} \qquad (2)$$

$$m \cdot v^2 = \frac{1}{4} \frac{Z \cdot e^2}{\pi \cdot \varepsilon_0 \cdot r} \qquad (3)$$

$$m = \frac{Z \cdot e^2}{4 \cdot \pi \cdot \varepsilon_0 \cdot v^2 \cdot r} \qquad (4)$$

Now, we write the known relation Lorenz (5), for the mass of a corpuscle in function of the corpuscle squared speed.

With the relations (4) and (5) one obtains the first essential expression (6).

$$m = \frac{m_0 \cdot c}{\sqrt{c^2 - v^2}} \qquad (5)$$

$$\frac{m_0 \cdot c}{\sqrt{c^2 - v^2}} = \frac{Z \cdot e^2}{4 \cdot \pi \cdot \varepsilon_0 \cdot v^2 \cdot r} \qquad (6)$$

One utilizes now, the Niels Bohr generalized relation (7).

It uses for the second time the Lorenz relation (5) with the Bohr relation (7) and in this mode one obtains the second essential expression (8).

$$m = \frac{n^2 \cdot \varepsilon_0 \cdot h^2}{\pi \cdot r \cdot e^2 \cdot Z} \qquad (7)$$

$$\frac{m_0 \cdot c}{\sqrt{c^2 - v^2}} = \frac{n^2 \cdot \varepsilon_0 \cdot h^2}{\pi \cdot r \cdot e^2 \cdot Z} \qquad (8)$$

Now, one keeps just the two essential expressions (6 and 8). It writes (8) in the form (8').

$$\sqrt{c^2 - v^2} \cdot n^2 \cdot \varepsilon_0 \cdot h^2 = \pi \cdot r \cdot m_0 \cdot c \cdot e^2 \cdot Z \qquad (8')$$

Elevating the relationship (8') to the square, to explicit the squared electron speed, it obtains the form (9).

$$v^2 = \frac{(n^4 \cdot \varepsilon_0^2 \cdot h^4 - \pi^2 \cdot r^2 \cdot m_0^2 \cdot e^4 \cdot Z^2) \cdot c^2}{n^4 \cdot \varepsilon_0^2 \cdot h^4} \qquad (9)$$

The formula (9) can be put in the form (10), where the constant k takes the form (10').

$$v^2 = c^2 - k \cdot c^2 \cdot r^2 \qquad (10)$$

$$k = \frac{\pi^2 \cdot m_0^2 \cdot e^4 \cdot Z^2}{n^4 \cdot \varepsilon_0^2 \cdot h^4} \qquad (10')$$

Now one writes the essential relation (6) in the form (6').

$$4 \cdot m_0 \cdot c \cdot \pi \cdot \varepsilon_0 \cdot r \cdot v^2 = Z \cdot e^2 \cdot \sqrt{c^2 - v^2} \qquad (6')$$

Then, putting the relation (6') at the square, it obtains the formula (6'').

$$16 \cdot m_0^2 \cdot c^2 \cdot \pi^2 \cdot \varepsilon_0^2 \cdot r^2 \cdot v^4 = Z^2 \cdot e^4 \cdot (c^2 - v^2) \quad (6'')$$

In the relation (6'') one introduce the squared velocity of the electron, taken from the expression (10) and one obtains the formula (11).

$$16 \cdot m_0^2 \cdot \pi^2 \cdot \varepsilon_0^2 \cdot (c^2 - k \cdot c^2 \cdot r^2)^2 = Z^2 \cdot e^4 \cdot k \quad (11)$$

The (11) relationship can be arranged in the form (12).

$$(c^2 - k \cdot c^2 \cdot r^2)^2 = \frac{Z^2 \cdot e^4 \cdot k}{16 \cdot m_0^2 \cdot \pi^2 \cdot \varepsilon_0^2} \quad (12)$$

One squares the relation (12) and it obtains the expression (13).

$$(c^2 - k \cdot c^2 \cdot r^2) = \pm \frac{Z \cdot e^2 \cdot \sqrt{k}}{4 \cdot m_0 \cdot \pi \cdot \varepsilon_0} \quad (13)$$

The relation (13) can be arranged to the form (14).

$$k \cdot c^2 \cdot r^2 = c^2 \mp \frac{Z \cdot e^2 \cdot \sqrt{k}}{4 \cdot m_0 \cdot \pi \cdot \varepsilon_0} \quad (14)$$

From relation (14) it explicit the squared electron radius and one obtains the relation (15).

$$r^2 = \frac{1}{k} \mp \frac{Z \cdot e^2}{4 \cdot m_0 \cdot \pi \cdot \varepsilon_0 \cdot \sqrt{k} \cdot c^2} \qquad (15)$$

Now, one exchange in the relation (15), the constant k with its expression (10') and it obtains the relation (16).

$$r^2 = \frac{n^4 \cdot \varepsilon_0^2 \cdot h^4}{\pi^2 \cdot m_0^2 \cdot e^4 \cdot Z^2} \mp \frac{n^2 \cdot h^2}{4 \cdot \pi^2 \cdot m_0^2 \cdot c^2} \qquad (16)$$

The expression (16) can be put in the form (17).

$$r^2 = \frac{n^4 \cdot \varepsilon_0^2 \cdot h^4}{\pi^2 \cdot m_0^2 \cdot e^4 \cdot Z^2} \cdot (1 \mp \frac{e^4 \cdot Z^2}{4 \cdot c^2 \cdot \varepsilon_0^2 \cdot h^2 \cdot n^2}) \qquad (17)$$

Extracting the square root of the expression (17), it obtains for the electron radius (r), the expression (18).

$$r = \pm \frac{n^2 \cdot \varepsilon_0 \cdot h^2}{\pi \cdot m_0 \cdot e^2 \cdot Z} \cdot \sqrt{1 \mp \frac{e^4 \cdot Z^2}{4 \cdot c^2 \cdot \varepsilon_0^2 \cdot h^2 \cdot n^2}} \qquad (18)$$

Physically there is only the positive solution (19).

$$r = + \frac{n^2 \cdot \varepsilon_0 \cdot h^2}{\pi \cdot m_0 \cdot e^2 \cdot Z} \cdot \sqrt{1 \mp \frac{e^4 \cdot Z^2}{4 \cdot c^2 \cdot \varepsilon_0^2 \cdot h^2 \cdot n^2}} \qquad (19)$$

The relation (19) is writing in final form (20) [3].

$$r = \frac{n^2 \cdot \varepsilon_0 \cdot h^2}{\pi \cdot m_0 \cdot e^2 \cdot Z} \cdot \sqrt{1 \mp \frac{e^4 \cdot^2}{4 \cdot c^2 \cdot \varepsilon_0^2 \cdot h^2 \cdot n^2}} \qquad (20)$$

The expression (20) it's not just a new theory for calculating the radius with that the electron is running around the nucleus of an atom, it is also a really new theory of an atomic model, or a new quantum theory.

For a value of the quantum number n (for a constant atomic number Z), we haven't just one energetically level (like in the Bohr model).

Now we can find two energetically below levels, which form an electronic layer, an electronic cloud. For example, for n=1, we have two sublevels (two below levels) [1-2].

USED NOTATIONS

The permissive constant (the permittivity):
$$\varepsilon_0 = 8.85418 \cdot 10^{-12} [\frac{C^2}{N \cdot m^2}];$$

The Planck constant:
$$h = 6.626 \cdot 10^{-34} [J \cdot s];$$

The rest mass of electron:
$$m_0 = 9.1091 \cdot 10^{-31} [kg];$$

The Pythagoras number:
$$\pi = 3.141592654;$$

The electrical elementary load:

$$e = -1.6021 \cdot 10^{-19}[C];$$

The light speed in vacuum:

$$c = 2.997925 \cdot 10^8 [\frac{m}{s}];$$

n=the principal quantum number (the Bohr quantum number);

Z=the number of protons from the atomic nucleus (the atomic number) [2].

DETERMINING THE TWO DIFFERENT ELECTRON SPEED VALUES

Relationship (6'') may be written in the form (6''') [2].

$$16 \cdot m_0^2 \cdot c^2 \cdot \pi^2 \cdot \varepsilon_0^2 \cdot r^2 \cdot v^4 + \\ + Z^2 \cdot e^4 \cdot v^2 - Z^2 \cdot e^4 \cdot c^2 = 0 \qquad (6''')$$

It can see easily that the relation (6''') represents a two degree equation in v^2.

One calculates v^2 with the formula (6^{IVa}).

$$v_{1,2}^2 = \frac{-Z^2 \cdot e^4 \pm \sqrt{Z^4 \cdot e^8 + 8^2 \cdot m_0^2 \cdot \pi^2 \cdot \varepsilon_0^2 \cdot c^4 \cdot Z^2 \cdot e^4 \cdot r^2}}{2 \cdot 16 \cdot m_0^2 \cdot c^2 \cdot \pi^2 \cdot \varepsilon_0^2 \cdot r^2} \quad (6^{\mathrm{IVa}})$$

Physically there is just the positive solution, and one keeps it for the relation (6^{IV}) (only the positive sign) [2].

$$v^2 = \frac{-Z^2 \cdot e^4 + \sqrt{Z^4 \cdot e^8 + 8^2 \cdot m_0^2 \cdot \pi^2 \cdot \varepsilon_0^2 \cdot c^4 \cdot Z^2 \cdot e^4 \cdot r^2}}{2 \cdot 16 \cdot m_0^2 \cdot c^2 \cdot \pi^2 \cdot \varepsilon_0^2 \cdot r^2} \quad (6^{\mathrm{IV}})$$

It can thinks that the relation (6^{IV}) gives only one solution for the electron squared speed (v^2), but really there is two solutions for this parameter, v^2, because the value of the squared radius (r^2) gives two physically solutions. It put the relation (6^{IV}) in the form (6^{V}) [2].

$$v_{1,2}^2 = \frac{-1 + \sqrt{1 + \dfrac{8^2 \cdot m_0^2 \cdot \pi^2 \cdot \varepsilon_0^2 \cdot c^2}{Z^2 \cdot e^4} \cdot c^2 \cdot r^2}}{\dfrac{1}{2} \cdot \dfrac{8^2 \cdot m_0^2 \cdot c^2 \cdot \pi^2 \cdot \varepsilon_0^2}{Z^2 \cdot e^4} \cdot r^2} \quad (6^{\mathrm{V}})$$

The formula (6^{V}) can be written in the form (6^{VI}), where the constant k_1 takes the form (6^{VII}) [2].

$$v_{1,2}^2 = \frac{\sqrt{1 + k_1 \cdot c^2 \cdot r^2} - 1}{\dfrac{k_1}{2} \cdot r^2} \quad (6^{\mathrm{VI}})$$

$$k_1 = \frac{8^2 \cdot m_0^2 \cdot \pi^2 \cdot \varepsilon_0^2 \cdot c^2}{Z^2 \cdot e^4} \qquad (6^{VII})$$

Now one starts with relation (6^{VI}) who can be written in the form (21).

$$v^2 = \frac{2 \cdot c^2}{\sqrt{1 + k_1 \cdot c^2 \cdot r^2} + 1} \qquad (21)$$

One notes the radical with R (see the relation 22).

$$R = \sqrt{1 + k_1 \cdot c^2 \cdot r^2} \qquad (22)$$

In relation (22) one introduces for r^2 the expression (20) and it obtains the form (22').

$$R = \sqrt{1 + \frac{k_1 \cdot c^2}{k} \cdot (1 \mp \frac{2 \cdot \sqrt{k}}{c \cdot \sqrt{k_1}})} \qquad (22')$$

In relation (22') one exchanges the two constant k_1 and k with the two values from expressions (6^{VII}) respective (10') and it obtains for (22') the form (22'') [2].

$$R = \sqrt{1 + \frac{8^2 m_0^2 \cdot \pi^2 \cdot \varepsilon_0^2 \cdot c^4 \cdot n^4 \cdot \varepsilon_0^2 \cdot h^4}{Z^2 \cdot e^4 \cdot \pi^2 \cdot m_0^2 \cdot e^4 \cdot Z^2} \cdot (1 \mp \frac{2\pi \cdot m_0 \cdot e^4 \cdot Z^2}{8n^2 \cdot \varepsilon_0^2 \cdot h^2 \cdot c^2})} \quad (22'')$$

One put the expression (22'') in the form (22''').

$$R = \sqrt{1 + \frac{8^2 \cdot \varepsilon_0^4 \cdot c^4 \cdot h^4 \cdot n^4}{e^8 \cdot Z^4} (1 \mp \frac{e^4 \cdot Z^2}{4\varepsilon_0^2 \cdot c^2 \cdot h^2 \cdot n^2})} \quad (22''')$$

The expression (22''') will be written in the form (22^{IV}).

$$R = \sqrt{1 + \frac{8^2 \cdot \varepsilon_0^4 \cdot c^4 \cdot h^4 \cdot n^4}{e^8 \cdot Z^4} \mp \frac{2 \cdot 8 \cdot \varepsilon_0^2 \cdot c^2 \cdot h^2 \cdot n^2}{e^4 \cdot Z^2}} \quad (22^{IV})$$

The expression (22^{IV}) can be restricted to the forms (22^V) and (22^{VI}).

$$R = \sqrt{\left(1 \mp \frac{8 \cdot \varepsilon_0^2 \cdot c^2 \cdot h^2 \cdot n^2}{e^4 \cdot Z^2}\right)^2} \quad (22^V)$$

$$R = \left| 1 \mp \frac{8 \cdot \varepsilon_0^2 \cdot c^2 \cdot h^2 \cdot n^2}{e^4 \cdot Z^2} \right| \quad (22^{VI})$$

One notes with E the expression (23).

$$E = \frac{8 \cdot \varepsilon_0^2 \cdot c^2 \cdot h^2}{e^4} \cdot \frac{n^2}{Z^2} \qquad (23)$$

This expression must be evaluated.

$$E = \frac{8 \cdot 8.85418^2 \cdot 10^{-24} \cdot 2.997925^2 \cdot 10^{16}}{1.6021^4 \cdot 10^{-76}} \cdot$$
$$\cdot \frac{6.626^2 \cdot 10^{-68} \cdot n^2}{Z^2} = \frac{37564.06551 \cdot n^2}{Z^2} \qquad (23')$$

For Zmax=92, we have a minimum of expression E (23''):

$$E_{min} = 4.438098477 \cdot n^2 \qquad (23'')$$

It can see easily that Emin > 1:

$$E_{min} \succ 1 \qquad (24)$$

Now, one can write the expression (22^{VI}) in the forms (22^{VII}) a, and b:

$$R_1 = E - 1 \qquad (22^{VIIa})$$

$$R_2 = E + 1 \qquad (22^{VIIb})$$

Only now the expression (21) can be evaluated and reduced to two forms (21^{Ia}) and respective (21^{Ib}):

$$v_1^2 = \frac{2 \cdot c^2}{E - 1 + 1} \qquad (21^{Ia})$$

$$v_2^2 = \frac{2 \cdot c^2}{E + 1 + 1} \qquad (21^{Ib})$$

The two relations take the forms (21^{II}) a, and b:

$$v_1^2 = \frac{c^2}{\dfrac{E}{2}} \qquad (21^{IIa})$$

$$v_2^2 = \frac{c^2}{\dfrac{E}{2} + 1} \qquad (21^{IIb})$$

If one replaces E with its expression (23) it obtains for the electron speeds the relations (21^{III}) a, and b [2].

$$v_1^2 = \frac{e^4 \cdot Z^2}{4 \cdot \varepsilon_0^2 \cdot h^2 \cdot n^2} \qquad (21^{IIIa})$$

$$v_2^2 = \frac{c^2}{\dfrac{4 \cdot \varepsilon_0^2 \cdot c^2 \cdot h^2 \cdot^2}{e^4 \cdot Z^2} + 1} \qquad (21^{IIIb})$$

DETERMINING THE MASSES AND THE ENERGY OF THE ATOMIC ELECTRON IN MOVEMENT

The exact squared speeds can be written in the forms (25, 26) [2].

$$r_- = r_1 \Rightarrow v_1^2 = \frac{e^4 \cdot Z^2 \cdot c^2}{4 \cdot \varepsilon_0^2 \cdot c^2 \cdot h^2 \cdot n^2} \qquad (25)$$

$$r_+ = r_2 \Rightarrow v_2^2 = \frac{e^4 \cdot Z^2 \cdot c^2}{4 \cdot \varepsilon_0^2 \cdot c^2 \cdot h^2 \cdot n^2 + e^4 \cdot Z^2} \qquad (26)$$

With these velocities one can write the two adequate masses (27), (28) [2].

$$r_- = r_1 \Rightarrow m_1 = \frac{m_0}{\sqrt{1 - \dfrac{e^4 \cdot Z^2}{4 \cdot \varepsilon_0^2 \cdot c^2 \cdot h^2 \cdot n^2}}} \qquad (27)$$

$$r_+ = r_2 \Rightarrow m_2 = \frac{m_0}{\sqrt{1 - \dfrac{e^4 \cdot Z^2}{4 \cdot \varepsilon_0^2 \cdot c^2 \cdot h^2 \cdot n^2 + e^4 \cdot Z^2}}} \qquad (28)$$

The total electron energy can be written in the forms (29) and (30) [2].

$$r_- = r_1 \Rightarrow W_1 = \frac{m_0 \cdot c^2}{\sqrt{1 - \dfrac{e^4 \cdot Z^2}{4 \cdot \varepsilon_0^2 \cdot c^2 \cdot h^2 \cdot n^2}}} \qquad (29)$$

$$r_+ = r_2 \Rightarrow W_2 = \frac{m_0 \cdot c^2}{\sqrt{1 - \dfrac{e^4 \cdot Z^2}{4 \cdot \varepsilon_0^2 \cdot c^2 \cdot h^2 \cdot n^2 + e^4 \cdot Z^2}}} \qquad (30)$$

The possible frequency of pumping, between the two near energetically below levels can be written in the form (31) [2].

$$v = \frac{W_1 - W_2}{h} = \frac{m_0 \cdot c^2}{h} \cdot$$

$$\cdot \left[\frac{1}{\sqrt{1 - \dfrac{e^4 \cdot Z^2}{4 \cdot \varepsilon_0^2 \cdot c^2 \cdot h^2 \cdot n^2}}} - \right.$$

$$\left. - \frac{1}{\sqrt{1 - \dfrac{e^4 \cdot Z^2}{4 \cdot \varepsilon_0^2 \cdot c^2 \cdot h^2 \cdot n^2 + e^4 \cdot Z^2}}} \right]$$

(31)

THE *POSSIBLE* LASER FREQUENCIES

In the table 1, one can see the possible LASER pumping frequencies (all in visible domain $4.34*10^{14} \div 6.97*10^{14}$ [Hz]), calculated for different principal quantum number n.

The possible L A S E R pumping frequencies Table 1

n	Z	[zH]ν	Element	n	Z	[zH]ν	Element
2	15	=5.54942E14	P		78	=4.43344E+14	Pt
	22	=5.072E14	Ti		79	=4.66537E+14	Au
3	23	=6.0598E14	V		80	=4.90629E+14	Hg
	29	=4.8452E+14	Cu		81	=5.15642E+14	Tl
	30	=5.54942E+14	Zn		82	=5.41601E+14	Pb
4	31	=6.32782E+14	Ga		83	=5.68529E+14	Bi
	36	=4.71283E+14	Kr		84	=5.96449E+14	Po
	37	=5.25911E+14	Rb		85	=6.25386E+14	At
	38	=5.8516E+14	Sr		86	=6.55364E+14	Rn
5	39	=6.49284E+14	Y	11	87	=6.86408E+14	Fr
	43	=4.6261E+14	Tc		85	=4.41451E+14	At
	44	=5.072E+14	Ru		86	=4.6261E+14	Rn
	45	=5.54942E+14	Rh		87	=4.8452E+14	Fr
	46	=6.0598E+14	Pd		88	=5.072E+14	Ra
6	47	=6.60463E+14	Ag		89	=5.30668E+14	Ac
	50	=4.56488E+14	Sn		90	=5.54942E+14	Th
	51	=4.94145E+14	Sb		91	=5.8004E+14	Pa
	52	=5.34086E+14	Te		92	=6.0598E+14	U
	53	=5.76403E+14	I		93	=6.32782E+14	Np
	54	=6.21189E+14	Xe		94	=6.60463E+14	Pu
7	55	=6.68536E+14	Cs	12	95	=6.89044E+14	Am

	57	=4.51937E+14	La		92	=4.39854E+14	U
	58	=4.8452E+14	Ce		93	=4.59306E+14	Np
	59	=5.18835E+14	Pr		94	=4.79396E+14	Pu
	60	=5.54942E+14	Nd		95	=5.00139E+14	Am
	61	=5.92904E+14	Pm		96	=5.21548E+14	Cm
	62	=6.32782E+14	Sm		97	=5.43638E+14	Bk
8	63	=6.7464E+14	Eu		98	=5.66422E+14	Cf
	64	=4.48422E+14	Gd		99	=5.89916E+14	Es
	65	=4.77132E+14	Tb		100	=6.14134E+14	Fm
	66	=5.072E+14	Dy		101	=6.39091E+14	Md
	67	=5.38669E+14	Ho		102	=6.64801E+14	No
	68	=5.71581E14	Er	13	103	=6.9128E+14	Lw
	69	=6.0598E+14	Tm		99	=4.38489E+14	Es
	70	=6.4191E+14	Yb		100	=4.56488E+14	Fm
9	71	=6.79416E+14	Lu		101	=4.75037E+14	Md
	71	=4.45624E+14	Lu		102	=4.94145E+14	No
	72	=4.71283E+14	Hf		103	=5.13824E+14	Lr
	73	=4.98035E+14	Ta		104	=5.34086E+14	Rf
	74	=5.25911E+14	W	14	105	=5.54942E+14	Db
	75	=5.54942E+14	Re				
	76	=5.8516E+14	Os				
	77	=6.16596E+14	Ir				
	78	=6.49284E+14	Pt				
10	79	=6.83255E+14	Au				

THE LASER FREQUENCIES AND CONCLUSIONS

If the second speed value does not exist physically, we must calculate the new atomic model just for the new first value, with the next relations:

$$r = \frac{n^2 \cdot \varepsilon_0 \cdot h^2}{\pi \cdot m_0 \cdot e^2 \cdot Z} \cdot \sqrt{1 - \frac{e^4 \cdot Z^2}{4 \cdot c^2 \cdot \varepsilon_0^2 \cdot h^2 \cdot n^2}} \qquad (20')$$

$$v^2 = \frac{e^4 \cdot Z^2}{4 \cdot \varepsilon_0^2 \cdot h^2 \cdot n^2} \qquad (25')$$

$$m = \frac{m_0}{\sqrt{1 - \frac{e^4 \cdot Z^2}{4 \cdot \varepsilon_0^2 \cdot c^2 \cdot h^2 \cdot n^2}}} \qquad (27')$$

$$W = \frac{m_0 \cdot c^2}{\sqrt{1 - \frac{e^4 \cdot Z^2}{4 \cdot \varepsilon_0^2 \cdot c^2 \cdot h^2 \cdot n^2}}} \qquad (29')$$

$$\gamma = \frac{m_0.c^2}{h} \left(\frac{1}{\sqrt{1 - \dfrac{e^4.Z^2}{4.\varepsilon_0^2.c^2.h^2.n_1^2}}} - \frac{1}{\sqrt{1 - \dfrac{e^4.Z^2}{4.\varepsilon_0^2.c^2.h^2.n_2^2}}} \right)$$

$$(31')$$

The pumping frequency required to achieve the transition of the electrons between two energetically levels can be written in the form (31').

In the table 2, one can see the LASER pumping frequencies.

All frequencies are outside visible area. One can make Ultraviolet Frequency-X ray LASER.

The bold value can be used to make a Rubin (Crystal) LASER.

The paper realizes a new atomic model and a new quantum theory (relation 20').

It determines as well the frequency of pumping for the transition between two energetically levels, with possible applications in LASER, MASER, IRASER industry (relation 31').

The pumping frequencies, between two nearer level								Table 2
Z	ν	El n_1-n_2	Z	ν	Element	Z	ν	Element
1		H	2		He	3	2.22122E+16	Li 1-2
4	3.95022E+16	Be 1-2	5	6.17499E+16	B 1-2	6	8.89688E+16	C 1-2
7	1.21175E+17	N 1-2	8	1.58388E+17	O 1-2	9	2.00631E+17	F 1-2
10	2.47929E+17	Ne 1-2	11	5.53738E+16	Na 2-3	12	6.59213E+16	Mg 2-3
13	7.73939E+16	Al 2-3	14	8.97936E+16	Si 2-3	15	1.03123E+17	P 2-3
16	1.17383E+17	S 2-3	17	1.32578E+17	Cl 2-3	18	1.48709E+17	Ar 2-3
19	5.7866E+16	K 3-4	20	6.41348E+16	Ca 3-4	21	7.07288E+16	Sc 3-4
22	7.76485E+16	Ti 3-4	23	8.48944E+16	V 3-4	24	**9,24672E+16**	Cr 3-4
25	1.00368E+17	Mn 3-4	26	1.08596E+17	Fe 3-4	27	1.17153E+17	Co 3-4
28	1.2604E+17	Ni 3-4	29	1.35258E+17	Cu 3-4	30	1.44806E+17	Zn 3-4
31	1.54686E+17	Ga 3-4	32	1.64899E+17	Ge 3-4	33	1.75446E+17	As 3-4
34	1.86327E+17	Se 3-4	35	1.97544E+17	Br 3-4	36	2.09097E+17	Kr 3-4
37	1.01887E+17	Rb 4-5	38	1.07502E+17	Sr 4-5	39	1.1327E+17	Y 4-5
40	1.19192E+17	Zr 4-5	41	1.25268E+17	Nb 4-5	42	1.31498E+17	Mo 4-5
43	1.37882E+17	Tc 4-5	44	1.44421E+17	Ru 4-5	45	1.51116E+17	Rh 4-5
46	1.57966E+17	Pd 4-5	47	1.64972E+17	Ag 4-5	48	1.72134E+17	Cd 4-5
49	1.79453E+17	In 4-5	50	1.86928E+17	Sn 4-5	51	1.94561E+17	Sb 4-5
52	2.02352E+17	Te 4-5	53	2.10301E+17	I 4-5	54	2.18408E+17	Xe 4-5
55	1.22612E+17	Cs 5-6	56	1.2715E+17	Ba 5-6	57	1.31772E+17	La 5-6
58	1.36479E+17	Ce 5-6	59	1.41271E+17	Pr 5-6	60	1.46147E+17	Nd 5-6
61	1.51109E+17	Pm 5-6	62	1.56157E+17	Sm 5-6	63	1.6129E+17	Eu 5-6
64	1.66508E+17	Gd 5-6	65	1.71813E+17	Tb 5-6	66	1.77203E+17	Dy 5-6
67	1.8268E+17	Ho 5-6	68	1.88243E+17	Er 5-6	69	1.93893E+17	Tm 5-6
70	1.9963E+17	Yb 5-6	71	2.05453E+17	Lu 5-6	72	2.11364E+17	Hf 5-6
73	2.17362E+17	Ta 5-6	74	2.23448E+17	W 5-6	75	2.29621E+17	Re 5-6
76	2.35883E+17	Os 5-6	77	2.42232E+17	Ir 5-6	78	2.4867E+17	Pt 5-6
79	2.55197E+17	Au 5-6	80	2.61813E+17	Hg 5-6	81	2.68517E+17	Tl 5-6
82	2.75311E+17	Pb 5-6	83	2.82195E+17	Bi 5-6	84	2.89168E+17	Po 5-6
85	2.96231E+17	At 5-6	86	3.03385E+17	Rn 5-6	87	1.8618E+17	Fr 6-7
88	1.90549E+17	Ra 6-7	89	1.94972E+17	Ac 6-7	90	1.99447E+17	Th 6-7
91	2.03976E+17	Pa 6-7	92	2.08557E+17	U 6-7	93	2.13193E+17	Np 6-7
94	2.17881E+17	Pu 6-7	95	2.22624E+17	Am 6-7	96	2.2742E+17	Cm 6-7
97	2.3227E+17	Bk 6-7	98	2.37174E+17	Cf 6-7	99	2.42131E+17	Es 6-7
100	2.47144E+17	Fm 6-7	101	2.5221E+17	Md 6-7	102	2.57331E+17	No 6-7
103	2.62506E+17	Lr 6-7	104	2.67736E+17	Rf 6-7	105	2.73021E+17	Db 6-7

BIBLIOGRAPHY

[1] David Halliday, Robert, R., - *Physics, Part II,* Edit. John Wiley & Sons, Inc. - New York, London, Sydney, 1966;

[2] Petrescu F.I., *The movement of an electron around the atomic nucleus,* in ICOME 2010, Craiova, 2010.

CHAPTER II – SOME FEW SPECIFICATIONS ABOUT THE DOPPLER EFFECT TO THE ELECTROMAGNETIC WAVES

Introduction

The Doppler Effect [1-3] represents the frequency variation of the waves, received by an observer which is drawing (coming), respectively it's removing (going), from a wave spring (source).

If a bright spring is drawing to an observer, the frequency of waves received by the observer is bigger than the emitted frequency of source, such that the respective spectral lines are moving to violet.

On the contrary, if the light source is removing from the observer, the spectral lines are moving to red.

One proposes to study the Doppler Effect for the light waves, generally for the electromagnetic waves.

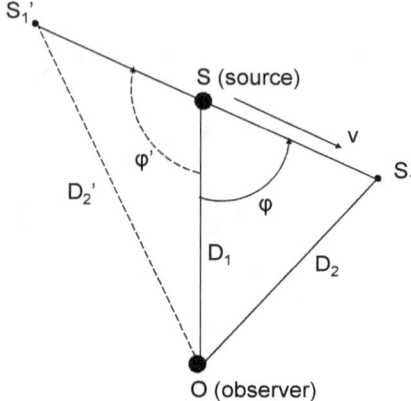

Fig. 1. *The waves received by an observer O from a waves source S, which is moving in relation with the observer, by the direction SS₁*

The new relations

We wish to calculate the period (T [s]) of the waves received by an observer O (figure 1) from a waves source S, which is moving in relation with the observer, on the direction SS_1 with the relative speed v [m/s] [1, 2].

T_0 [s] is the period of waves emitted by the source S.

At the moment t_0 [s], determinate by the observer O, from the source S bend a bright wave; this wave traverse the distance $D_1=SO$ [m] and arrive in O at the moment t_1 [s].

$$t_1 = t_0 + \frac{D_1}{c} \qquad (1)$$

where c is the light speed in vacuum: $c \cong 3.10^8$ [m/s].

After a T_0 period, from the source S (arrived now in S_1), from the source S_1 starts a second wave. The distance SS_1 [m] is:

$$SS_1 = v.T_0 \tag{2}$$

The observer O, receive the second waves at the moment t_2 [s]:

$$t_2 = t_0 + T_0 + \frac{D_2}{c} \tag{3}$$

The period T is equal with the difference between the two moments.

$$T = t_2 - t_1 = T_0 + \frac{D_2 - D_1}{c} \tag{4}$$

The angle φ [rad] between the two vectors, SS_1 and SO is known and the distance $D_1 = SO$ is known as well. With the COS theorem in the certain triangle SOS_1 one obtains the distance D_2 [m]:

$$D_2 = \sqrt{D_1^2 + SS_1^2 - 2.D_1.SS_1.\cos\varphi} \tag{5}$$

With SS_1 from (2) the relation (5), become the expression (6).

$$D_2 = \sqrt{D_1^2 + v^2.T_0^2 - 2.D_1.v.T_0.\cos\varphi} \qquad (6)$$

With the expression (6) in relation (4) one obtains the form (7).

$$T = T_0 + \frac{\sqrt{D_1^2 + v^2 T_0^2 - 2D_1 v T_0 \cos\varphi} - D_1}{c} \qquad (7)$$

The relation (7) can be put in the form (8).

$$T = T_0 (1 + \beta \frac{v.T_0 - 2.D_1.\cos\varphi}{\sqrt{D_1^2 + v^2 T_0^2 - 2D_1 v T_0 \cos\varphi} + D_1}) \qquad (8)$$

where β is the ratio between the two speed, v and c:

$$\beta = \frac{v}{c} \qquad (9)$$

Presents the classical relation (10)

The classical relation (10) is very simply, but it's an approximate relation [2-3].

The expression (8) is more difficult but it's a very exact relation. It can be put in the forms (18), (19) and finally (20).

$$\frac{T}{T_0} = 1 \pm \beta.\cos\varphi \qquad (10)$$

Some aspects

a) When the source S is removing from the observer, the angle φ (see the figure 1) take the values (φ') comprised between 90^0 and 180^0, $\cos\varphi$ become negative, the numerator of expression (8) become positive and the period of observer O (T) it'll be always bigger than T_0 (the period of source): $T > T_0$ and $y < y_0$ (the spectral lines are red).

When the source S is drawing to the observer, the angle $\varphi \in [0^0, 90^0)$ and $\cos\varphi > 0$. In this case one analyzes (11) the numerator of expression (8) and one can have two case (b and c) [1]:

$$N = v.T_0 - 2.D_1.\cos\varphi \qquad (11)$$

b) If $N < 0$, then $v.T_0 < 2.D_1.\cos\varphi$ or

$$\cos\varphi > \frac{v.T_0}{2.D_1} \qquad (12)$$

and $T<T_0$, or $y>y_0$ (the spectral lines are violet) [1].

c) If N>0, then

$$\cos\varphi < \frac{v.T_0}{2.D_1} \qquad (13)$$

and $T>T_0$, or $y<y_0$ (the spectral lines are red).

This case it wasn't known by the classical expression (10) [1].

d) The most interesting case is then when the angle $\varphi=90^0$, and $\cos\varphi=0$, when the source is moving perpendicular at the axle SO (see the figure 2). In this case the relation (8), become the expression (14).

$$T = T_0\left(1 + \frac{\beta.v.T_0}{\sqrt{D_1^2 + v^2.T_0^2} + D_1}\right) \qquad (14)$$

$T>T_0$ and $y<y_0$ (the spectral lines are red) [1].

This fact can't be seen by the classical relation (10) which (for the $\varphi=90^0$), takes the form (15):

$$T = T_0 \qquad (15)$$

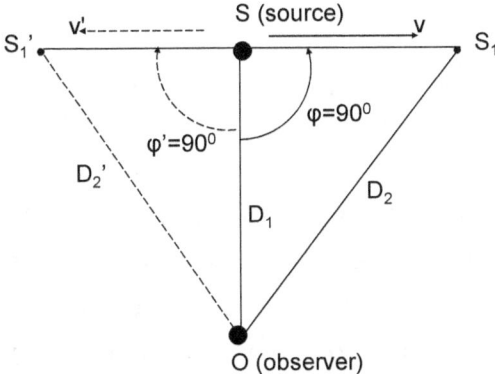

Fig. 2. *The waves received by an observer O from a waves source S when the source is moving perpendicular at the axle SO (it is a particular situation)*

The classical approximate relation (10, form 15) can't foresee the Doppler Effect for this case, but the effect virtually exist, and for this reason it was introduced the relativity effect (or the Lorentz transformation), where the period T_0 takes the form T_0/α (see [1]), and the relation (15) takes the form (16) [2, 3]:

$$T = \frac{T_0}{\alpha} \qquad (16)$$

where α is:

$$\alpha = \sqrt{1 - \beta^2} \qquad (17)$$

If v<c, the expression

$$\sqrt{D_1^2 + v^2 \cdot T_0^2 - 2 \cdot D_1 \cdot v \cdot T_0 \cdot \cos\varphi} \to D$$ and the relation
(8) can be approximated by the expression (18), (8=>18):

$$\frac{\gamma_0}{\gamma} = \frac{T}{T_0} = 1 - \beta \cdot \cos\varphi + \beta \cdot \frac{v \cdot T_0}{2 \cdot D_1} \qquad (18)$$

The distance D (D_1) can take different values for the same frequency γ_0 (One can't determine D from 8 or 18; D is indeterminate. Practically, the frequency γ is a real function of γ_0 and β; γ is a function of γ_0, T_0, or $\lambda_0 = c \cdot T_0$; The distance D can't takes any value; It must be a multiple of λ_0). The relation (18) takes mandatory the forms (19) for a quantum distance ($D_1 = n \cdot c \cdot T_0$) and (20) when n takes mandatory the basic value (n=1) to keep the own original wave (one utilize just the basic frequency for n=1, see the final relation 20; for other frequencies then we can already speak about other waves):

$$\frac{\gamma_0}{\gamma} = \frac{T}{T_0} = 1 - \beta \cdot \cos\varphi + \frac{1}{2} \cdot \beta^2 \cdot \frac{1}{n} \qquad (19)$$

$$\frac{\gamma_0}{\gamma} = \frac{T}{T_0} = 1 - \beta \cdot \cos\varphi + \frac{1}{2} \cdot \beta^2 \qquad (20)$$

First, the relation (20) can be utilized to determine the period T when one know the source period T_0 and the source velocity, v ($\beta = \dfrac{v}{c}$); one can speak now about a quantum Doppler Effect relation (20).

Second, if one know the two frequencies (γ, γ_0), one can determine the source velocity v in relation of the observer (β and v=β.c), with the new relation (20) or more rapidly with the classical form (10).

Conclusion

In this paper one proposes to exchange the classical relation (10) (see [1], p. 114) with the new and more exactly relation (20).

Bibliography

[1] **Bărbulescu N.**, *"Bazele fizice ale relativității Einsteiniene"*. Editura Ştiinţifică şi Enciclopedică, Bucureşti, 1979, p. 142-148;

[2] **David Halliday, Robert, R.**, - *Physics, Part II,* Edit. John Wiley & Sons, Inc. - New York, London, Sydney, 1966;

[3] **Petrescu-Prahova, M., Petrescu-Prahova, I.**, - *Fizica-Manual pentru anul IV liceu, secţia reală,* Editura Didactică şi Pedagogică, Bucureşti, 1976.

CHAPTER III - The Future Energy

Introduction

In the years 70-80 (1970-1980) it foreshadows a serious energy crisis with the rapid depletion of known reserves of oil and gas. The consequences would be catastrophic for mankind, but fortunately came just in time energy produced by nuclear fission. With Nuclear power we have saved, so they were a necessary evil.

Another 2-3 cycles (a cycle is about 30-40 years) they could be useful (even if they will evolve and will use the energy produced by fusion, in which case their effectiveness will increase considerably). However, must to prepare from time, the new energy of the future, the mankind future energy.

The most elegant solution which can be now seen is solar energy. This is practically inexhaustible, in quantities much greater than the planet needs, it is the clean, handy and can become the most affordable (if panels with photovoltaic cells will be produced in industrial quantities increasing).

For make this method to obtaining solar energy, to be totally clean, it is absolutely imperative that converted solar energy into electricity to be distributed directly to national energy networks, to avoid the use of different batteries (polluting).

Although it is small, efficiency of energy conversion (in cells), has increased and will increase further due to scientific research in the field.

It must be made the indication that, all living matter on Earth is the energy of the sun, either directly or indirectly.

The discovery of real particles faster than the speed of light (probably smaller than those known today) may open new chapters in the development of mankind, first in the energy field. Going deeper into matter, and by passing from the quantum level to the sub quantum level, or maybe even deeper, it will determine the increasing energy.

Matter is structured in such a way as if we can penetrating more inside it, the particles of which is composed are increasingly smaller and lighter, more dynamic and more energetic.

Although particle mass decreases, the speed is much greater, so the particle energy is much higher (the energy increases with the mass and with the velocity squared of the particle).

Links to quantum levels (within the atom) are more powerful than the chemical-molecular (in the molecule or between atoms), but lower than those of sub quantum level (in the atomic nucleus, between nucleons), which in turn are overshadowed by the level immediately below, the sub-sub quantum level (in nucleon, between the particles that compose it), and so on until we reach the basic level at which the particle can no longer be divided into other components. If the binding energy is higher, the energy released or required to break or compose these connections, is greater as well.

Hydrogen, as a key component, can be obtained in multiple ways, from almost any item, by nuclear reactions, by the decomposition of water under the action of radiation, by electrolysis of water, etc.

Burning hydrogen it is not a real source of energy (as on Earth, the hydrogen element is not found so much in isolated forms that can be extracted directly and then used as fuel; hydrogen element is generally achieved with energy consumption greater than the energy released by burning it); but it is more a strategic fuel, like a fuel which can be the life-long of internal combustion engines when the oil fuel will lessen or even will disappear.

Wind energy does not represent a real alternative energy, but in some cases it may be a component to complete certain energy goals.

The energy produced from thermal springs in some areas of the planet is very useful, but are very little compared with the needs of the earth.

Probably wave energy of seas and oceans has not given good results since it was not extended and imposed, the more so as we have a planet covered with water at the rate of 70%.

Maybe in the future, the man will exploit the temperature difference between the different levels of seas and oceans, to produce such energy (energies from seas water).

For now, the water remains a serious source of energy in the chapter, hydropower. From water, it extracts the hydrogen, which through burning turns back into water. From water it obtains "heavy water" (by the converting of the element Hydrogen, into heavy isotope named Deuterium, which contains in nucleus in addition to a proton and a neutron), and which is used as nuclear fuel, in some nuclear power plant.

If we look, retrospect and global, the water and the sun are the major energy sources of our planet. Even the living matter (including man), represents a very high proportion water. The water intervenes directly or indirectly in several ways, into the cellular level processes.

Obtaining Energy by the Annihilation of an Electron with a Positron, or Annihilation of a Proton with an Antiproton (case studies presentation)

Getting energy, renewable, clean, friendly (not dangerous), cheaper, by annihilation (For example, the annihilation of an electron with an anti electron). Electron and positron are obtained by extracting them from atoms; the extraction, consume a negligible amount of energy.

Then, the two particles are brought near one another (collision); now it occur the phenomenon of annihilation, when the rest mass is converted totally into energy (gamma photons).

Occur gamma photons, as many as needed to retrieve the total energy of the electron and positron (rest energy and kinetic energy); usually one can get two or three gamma particles (when we have a lower annihilation, ie two antiparticles with lower energy, each with a little beyond rest mass, ie the particles are accelerated at a low-speed motion), but we can get more particles when we have a high annihilation (ie when the particle energy is high and the particles were strongly accelerated before the collision).

Rest energy of an electron-positron pairs exceeds slightly 1 MeV (what is an extremely large energy from some as small particles, comparable energy with that achieved by the merger of two much larger particles, having rest mass of about 2000 times higher).

Hence the first great advantage of the new method proposed, namely that if the most complex physical phenomenon so far tried to get inside the material energy (hot or cold fusion), draw only about a thousandth part of the rest mass of the particle, resulting in the fusion of two particles practically only the energy gap between energy particles being free and their energy when they are united, the proposed method to extract virtually all the internal energy of the particles annihilated.

We started with the electron positron pair because these small particles are more easily extracted from the atoms (the atoms are then immediately regenerated naturally, which determines the nature of renewable energy from the annihilation of particles).

Next step is to test the annihilation between a proton and an antiproton, because their mass is about 1800 times higher than that of the electron and positron, resulting in their annihilation as an energy by about 1000 times higher, ie instead of 1 MeV, 1 GeV (is considered as the only real obtained energy, the energy donated by the proton of the hydrogen ion; but the energy of an antiproton is considered to be donated by us almost entirely, for now, because to obtain today an antiproton we must accelerate some particles at very high-energy and then collide them).

So the real comparison must to be made between the deuterons fusion and annihilation process of a hydrogen ion (proton) with an antiproton. It will be a difference of energy of about 1000 times higher per pair of particles used, in favor of the annihilation process.

Practically it realizes the dream of extracting energy from all the matter. Another great advantage of this method is that no radioactive substances and are not radioactive wastes from the process. From this process we obtain only gamma photons (ie energy) and possibly other energetic mini particles. The process does not pose any threat to humans and the environment.

The energy produced is clean. The technology required is much simpler than nuclear (fission or fusion), cheaper and easier to maintain. Enough energy is given by the annihilation process (virtually unlimited), cheap, clean, safe, renewable immediately (sustainable), with technology made simple.

We can extract the energy of the rest mass of an electron. For a pair of an electron and a positron this energy is circa 1 MeV.

The "synchrotron radiation (synchrotron light source)" produces deliberated a radiation source. Electrons are accelerated to high speeds in several stages to achieve a final energy (that is typically in the GeV range). We need two synchrotrons, a synchrotron for electrons and another who accelerates positrons. The particles must to be collided, after they are being accelerated to an optimal energy level. All the energies are collected at the exit of the Synchrotrons, after the collision of the opposite particles. We will recover the accelerating energy, and in addition we also collect the rest energy of the electrons and positrons.

At a rate of 10^19 electrons/s we obtain an energy of about 7 GWh / year, if even are produced only half of the possible collisions. This high rate can be obtained with 60 pulses per minute and 10^19 electrons per pulse, or with 600 pulses per minute and 10^18 electrons per pulse. If we increase the flow rate of 1,000 times, we can have a power of about 7 TWh / year. This type of energy can be a complement of the fusion energy, and together they must replace the energy obtained by burning hydrocarbons.

Advantages of the annihilation of an electron with a positron, compared with the nuclear fission reactors, are disposal of radioactive waste, of the risk of explosion and of the chain reaction.

Energy from the rest mass of the electron is more easily controlled compared with the fusion reaction, cold or hot.

Now, we don't need of enriched radioactive fuel (as in nuclear fission case), by deuterium, lithium and of accelerated neutrons (like in the cold fusion), of huge temperatures and pressures (as in the hot fusion), etc.

Results and Discussion

How much energy, can we get from inside of the matter? Einstein has showed that from one kg of matter we could get the energy needs for entire Earth for a year:

$$E=m'c^2=1[\text{kg}]'(3'10^8)^2[(\text{m/s})^2]=9'10^{16}[\text{j}]=2,5'10^{10}[\text{KWh}]=2,5'10^7[\text{MWh}]=2,5'10^4[\text{GWh}]=25[\text{TWh}]$$

We could do this, but only if we could extract all the energy from inside the matter.

Through nuclear fusion reaction can be extracted only a part of the rest energy of the particles used. This drop of energy (1 / 1000 of the mass energy of a proton-neutron pairs) is called, discrepancy.

For a kg of particles proton-neutron pairs, fusion energy is about a thousand times smaller than the total energy of a kilogram of matter (only 29 [GWh] from the total internal energy, 25 [TWh]); and considering that a return of 100% fusion reaction, which can't be done anyway.

Theoretically speaking, we can't draw from within the matter (through nuclear fusion reaction) than at most the thousandth part of its energy. Having in view the yield of the nuclear fusion reaction, this obtained energy is and less.

Through reaction of nuclear fission, the energies obtained will be even smaller.

The solution proposed in this work, obtaining energy by the mutual annihilation of two opposite particles, makes possible the requirement of extracting whole energy contained in matter.

A pair formed by a particle and its antiparticle, are brought side by side, at a distance which allow the process of reciprocal annihilation.

To increase the yield of the annihilation reaction (the number of annihilated particles from all particles that exist), we can accelerate the particles and antiparticles separately, and then we may send them into a room where they encounter annihilation at speeds and energies high, or at velocities and energies very high.

If we use electrons and positrons for the reaction of annihilation, it results photons of the gamma type.

In this case, to prevent the possible decay of the obtained photons, again into electrons and positrons (for beginning of this annihilation process with success), the antiparticles and particles used in the process of annihilation, should be collided at low speeds and with low energy.

We can test then the optimum energy particle which permits the reaction with the maxim yield. It is necessary that most particles and antiparticles used, to meet and annihilate each other, and it should be stable as many of the obtained gamma particles.

Conclusions

The fission energy was a necessary evil. In this mode it stretched the oil life, avoiding an energy crisis. Even so, the energy obtained from hydrocarbons represents today about 66% of all energy used. At this rate of use of oil, it will be consumed in about 40 years. Today, the production of energy obtained by nuclear fusion is not yet perfect prepared. But time passes quickly. We must rush to implement of the additional sources of energy already known, but and find new energy sources. In these conditions the proposed method to obtaining energy by annihilation of matter and antimatter, can be a real alternative sources of renewable energy.

References:

[1] EWEA Executive summary "Analysis of Wind Energy in the EU-25" (PDF). European Wind Energy Association. http://www.ewea.org/fileadmin/ewea_documents/documents/publications/WETF/Facts_Summary.pdf EWEA Executive summary. Retrieved 2007-03-11.

[2] Massachusetts Institute of Technology (2010, September 13). Funneling solar energy: Antenna made of carbon nanotubes could make photovoltaic cells more efficient. *Science Daily*. Retrieved September 21, 2010, from http://www.sciencedaily.com/releases/2010/09/100912151548.htm

[3] "Towards Sustainable Production and Use of Resources: Assessing Biofuels". United Nations Environment Programme. 2009-10-16. http://www.unep.fr/scp/rpanel/pdf/Assessing_Biofuels_Full_Report.pdf. Retrieved 2009-10-24.

[4] Petrescu, F. New Aircraft. COMEC 2009, Braşov, ROMANIA, 2009.

CHAPTER IV - NEW AIRCRAFT

4.1. ION THRUSTER [1]

4.1.1. About the ion thruster

Speaking about a new ionic engine means to speak about a new aircraft.

The chapter presents shortly the actual ionic engines (called ion thrusters) and the new ionic (pulse) engines proposed by the author.

Ionic engine (ion thruster, which accelerates the positive ions through a potential difference) is about 10 times more effective than classic system based on combustion.

We can still improve the efficiency of 10-50 times if one uses pulses of positive ions accelerated in a cyclotron mounted on the ship; the efficiency can easily grow for 1000 times if the positive ions will be accelerated in a high energy synchrotron, synchrocyclotron or isochronous cyclotron (1-100 GeV). In this, the big classic synchrotron is reduced to a ring surface (magnetic core).

Future (ionic) engine will have mandatory a circular particle accelerator (high or very high energy).

We can thus increase the speed and autonomy of the ship using a less quantity of fuel and power.

One can use synchrotron radiation (synchrotron light, high intensity beams), like high intensity (X-ray or Gamma ray) radiation, as well. In this case will be a beam engine (not an ionic engine), it'll use only the power (energy, which can be solar energy, nuclear energy, or both) and so we will remove the fuel.

It proposes using a powerful LINAC at the exit of synchrotron (especially when one accelerates electrons) to not lose energy by photons premature emission.

With a new ionic engine one builds a new aircraft, which can travel through water and. This new aircraft will can accelerate directly, without an additional combustion engine and without gravity assists from other planets.

An *ion thruster* is a form of electric propulsion used for spacecraft propulsion that creates thrust by accelerating ions. Ion thrusters are characterized by how they accelerate the ions, using either electrostatic or electromagnetic force.

Electrostatic ion thrusters use the Coulomb Force and accelerate the ions in the direction of the electric field. Electromagnetic ion thrusters use the Lorentz Force to accelerate the ions. Note that the term "ion thruster" frequently denotes the electrostatic or gridded ion thrusters, only.

The thrust created in ion thrusters is very small compared to conventional chemical rockets, but a very high specific impulse, or propellant efficiency, is obtained.

Due to their relatively high power needs, given the specific power of power supplies, and the requirement of an environment void of other ionized particles, ion thrust propulsion currently is only practicable in outer space.

The first experiments with ion thrusters were carried out by Robert Goddard at Clark College from 1916-1917. The technique was recommended for near-vacuum conditions at high altitude, but thrust was demonstrated with ionized air streams at atmospheric pressure. The idea appeared again in Hermann Oberth's "Wege zur Raumschiffahrt" (Ways to Spaceflight), published in 1923.

A working ion thruster was built by Harold R. Kaufman in 1959 at the NASA Glenn facilities. It was similar to the general design of a gridded electrostatic ion thruster with mercury as its fuel. Suborbital tests of the engine followed during the 1960s and in 1964 the engine was sent into a suborbital flight aboard the Space Electric Rocket Test 1 (SERT 1). It successfully operated for the planned 31 minutes before falling back to Earth.

4.1.2. Hall effect thruster

The Hall effect thruster was studied independently in the U.S. and the USSR in the 1950s and 60s. However, the concept of a Hall thruster was only developed into an efficient propulsion device in the former Soviet Union, whereas in the U.S., scientists focused instead on developing gridded ion thrusters.

Hall effect thrusters were operated on Soviet satellites since 1972. Until the 1990s they were mainly used for satellite stabilization in North-South and in East-West directions. Some 100-200 engines completed their mission on Soviet and Russian satellites until the late

1990s. Soviet thruster design was introduced to the West in 1992 after a team of electric propulsion specialists, under the support of the Ballistic Missile Defense Organization, visited Soviet laboratories.

Ion thrusters utilize beams of ions (electrically charged atoms or molecules) to create thrust in accordance with Newton's third law. The method of accelerating the ions varies, but all designs take advantage of the charge/mass ratio of the ions. This ratio means that relatively small potential differences can create very high exhaust velocities. This reduces the amount of reaction mass or fuel required, but increases the amount of specific power required compared to chemical rockets. Ion thrusters are therefore able to achieve extremely high specific impulses. The drawback of the low thrust is low spacecraft acceleration because the mass of current electric power units is directly correlated with the amount of power given. This low thrust makes ion thrusters unsuited for launching spacecraft into orbit, but they are ideal for in-space propulsion applications.

Hall effect thrusters accelerate ions with the use of an electric potential maintained between a cylindrical anode and a negatively charged plasma which forms the cathode. The bulk of the propellant (typically xenon or bismuth gas) is introduced near the anode, where it becomes ionized, and the ions are attracted towards the cathode, they accelerate towards and through it, picking up electrons as they leave to neutralize the beam and leave the thruster at high velocity.

The anode is at one end of a cylindrical tube, and in the center is a spike which is wound to produce a radial magnetic field between it and the surrounding tube. The ions are largely unaffected by the magnetic field, since they are too massive. However, the electrons produced near the end of the spike to create the cathode are far more affected and are trapped by the magnetic field, and held in place by their attraction to the anode. Some of the electrons spiral down towards the anode, circulating around the spike in a Hall current. When they reach the anode they impact the uncharged propellant and cause it to be ionized, before finally reaching the anode and closing the circuit.

4.1.3. Gridded electrostatic ion thrusters

Gridded electrostatic ion thrusters commonly utilize xenon gas. This gas has no charge and is ionized by bombarding it with energetic electrons. These electrons can be provided from a hot cathode filament and accelerated in the electrical field of the cathode fall to the anode (Kaufman type ion thruster). Alternatively, the electrons can be accelerated by the oscillating electric field induced by an alternating magnetic field of a coil, which results in a self-sustaining discharge and omits any cathode (radiofrequency ion thruster).

The positively charged ions are extracted by an extraction system consisting of 2 or 3 multi-aperture grids. After entering the grid system via the plasma sheath the ions are accelerated due to the potential difference between the first and second grid (named screen and accelerator grid) to the final ion energy of typically 1-2 keV, thereby generating the thrust.

Ion thrusters emit a beam of positive charged xenon ions only. In order to avoid the charging-up of the spacecraft another cathode, placed near the engine, emits additional electrons (basically the electron current is the same as the ion current) into the ion beam. This also prevents the beam of ions from returning to the spacecraft and thereby cancelling the thrust.

Gridded electrostatic ion thruster research (past/present):

NASA Solar electric propulsion Technology Application Readiness (NSTAR)

NASA's Evolutionary Xenon Thruster (NEXT)

Nuclear Electric Xenon Ion System (NEXIS)

High Power Electric Propulsion (HiPEP)

EADS Radio-Frequency Ion Thruster (RIT)

Dual-Stage 4-Grid (DS4G)

4.1.4. Field Emission Electric Propulsion

Field Emission Electric Propulsion (FEEP) thrusters use a very simple system of accelerating liquid metal ions to create thrust. Most designs use either cesium or indium as the propellant. The design consists of a small propellant reservoir that stores the liquid metal, a very small slit that the liquid flows through, and then the accelerator ring.

Cesium and indium are used due to their high atomic weights, low ionization potentials, and low melting points. Once the liquid metal reaches the inside of the slit in the emitter, an electric field applied between the emitter and the accelerator ring causes the liquid metal to become unstable and ionize.

This creates a positive ion, which can then be accelerated in the electric field created by the emitter and the accelerator ring. These positively charged ions are then neutralized by an external source of electrons in order to prevent charging of the spacecraft hull.

4.1.5. Pulsed Inductive Thrusters

Pulsed Inductive Thrusters (PIT) use pulses of thrust instead of one continuous thrust, and have the ability to run on power levels in the order of Megawatts (MW).

PITs consist of a large coil encircling a cone shaped tube that emits the propellant gas. Ammonia is the gas commonly used in PIT engines.

For each pulse of thrust the PIT gives, a large charge first builds up in a group of capacitors behind the coil and is then released. This creates a current that moves circularly. The current then creates a magnetic field in the outward radial direction (Br), which then creates a current in the ammonia gas that has just been released in the opposite direction of the original current.

This opposite current ionizes the ammonia and these positively charged ions are accelerated away from the PIT engine due to the electric field crossing with the magnetic field Br, which is due to the Lorentz Force.

4.1.6. Magnetoplasmadynamic

Magnetoplasmadynamic (MPD) thrusters and Lithium Lorentz Force Accelerator (LiLFA) thrusters use roughly the same idea with the LiLFA thruster building off of the MPD thruster.

Hydrogen, argon, ammonia, and nitrogen gas can be used as propellant. The gas first enters the main chamber where it is ionized into plasma by the electric field between the anode and the cathode. This plasma then conducts electricity between the anode and the cathode.

This new current creates a magnetic field around the cathode which crosses with the electric field, thereby accelerating the plasma due to the Lorentz Force. The LiLFA thruster uses the same general idea as the MPD thruster, except for two main differences.

The first difference is that the LiLFA uses lithium vapor, which has the advantage of being able to be stored as a solid.

The other difference is that the cathode is replaced by multiple smaller cathode rods packed into a hollow cathode tube.

The cathode in the MPD thruster is easily corroded due to constant contact with the plasma. In the LiLFA thruster the lithium vapor is injected into the hollow cathode and is not ionized to its plasma form/corrode the cathode rods until it exits the tube.

The plasma is then accelerated using the same Lorentz Force.

4.1.7. Electrodeless Plasma Thrusters

Electrodeless Plasma Thrusters have two unique features, the removal of the anode and cathode electrodes and the ability to throttle the engine.

The removal of the electrodes takes away the factor of erosion which limits lifetime on other ion engines. Neutral gas is first ionized by electromagnetic waves and then transferred to another chamber where it is accelerated by an oscillating electric and magnetic field, also known as the ponderomotive force.

This separation of the ionization and acceleration stage give at the engine the ability to throttle the speed of propellant flow, which then changes the thrust magnitude and specific impulse values [1].

4.1.8. Plasma Micro Thruster

In the picture number 1 one presents „A Plasma Micro Thruster" Schematic and Prototype (see Figure 1, and [2]).

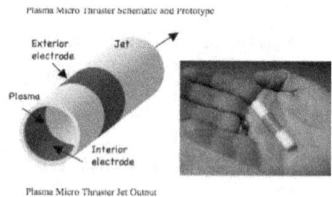

Fig. 1: *Plasma Micro Thruster, Schematic and Prototype*

4.2. THE HiPEP ENGINE

4.2.1. Powerful ion engine relies on microwaves

A powerful new ion propulsion system has been successfully ground-tested by NASA. The High Power Electric Propulsion ion engine trial marks the "first measurable milestone" for the ambitious $3 billion Project Prometheus, says director Alan Newhouse.

The HiPEP engine is the first tested propulsion technology with the potential power and longevity to thrust spacecraft as far as Jupiter without gravity assists from other planets.

These assists involve slingshot maneuvers around planets and can boost the speed of craft significantly. But they require specific planetary alignments, meaning suitable launch dates are rare.

In contrast, a probe powered by a HiPEP engine could launch any time. One goal of Project Prometheus, formerly called the Nuclear Systems Initiative, is to launch a spacecraft towards Jupiter by 2011. The flight would take at least eight years.

The key elements of the HiPEP engine are a high exhaust velocity, a microwave-based method for producing ions that performs for longer than existing technologies and a rectangular design that can more easily be scaled up than circular ones.

Spacecraft are increasingly being built with ion engines rather than engines that burn rocket fuel. This is because ion engines produce more power for a given amount of propellant, and provide a smooth output rather than intermittent spurts.

"Jupiter is such a far away target. Using a chemical system, you just couldn't do it," says John Foster, one of the principal creators of the engine at NASA's Glenn Research Center in Cleveland, Ohio.

The HiPEP engine differs from earlier ion engines, such as that powering NASA's Deep Space One mission, because the xenon ions are produced using a combination of microwaves and spinning magnets. Previously the electrons required were provided by a cathode. Using microwaves significantly reduces the wear and tear on the engine by avoiding any contact between the speeding ions and the electron source.

4.2.2. Nuclear fission

A Japanese asteroid-chasing spacecraft is already using microwave-based technology to produce ions, but Hayabusa uses a small device that could not produce enough power to fly to Jupiter. The HiPEP engine is currently capable of 12 kilowatts of power but its output will be boosted to at least 50 kW for the Jupiter mission.

The rectangular cross section of the HiPEP engine will make this easier, as it can be expanded along one of its sides. A circular engine would have to be rebuilt, says NASA.

Nonetheless, other researchers at NASA's Jet Propulsion Laboratory in Pasadena, California, are working on a cylindrical high-power ion engine, also for the Prometheus project. But Newhouse notes that building a powerful, long-lasting propulsion system is just "one of the pieces we need to get to Jupiter". The electricity for the ion engine is slated to come from on-board nuclear fission reactor. This part of the Prometheus Project is just beginning, with safety considerations, the miniaturization of the reactor and the identity of the fuel all needing to be decided.

4.3. NEW IONIC OR BEAM PULSES ENGINES

By this chapter the author proposes a new pulse engine which works with beam or ionic (ionic beam) pulses.

With a new ionic engine one builds a new aircraft (a new ship). The principal characteristic of this kind of engine is the high power (energy) which accelerates the beam at very high energy, in circular accelerators, in modern linear accelerators (LINAC), or in both.

One can use accelerators similar with the static physics accelerators (synchrotron, synchrocyclotron or isochronous cyclotron).

Ionic engine (ion thruster, which accelerates the positive ions through a potential difference) is about 10 times more effective than classic system based on combustion.

We can still improve the efficiency of 10-50 times if one uses positive ions accelerated in a cyclotron mounted on the ship; the efficiency can easily grow for 1000 times if the positive ions will be accelerated in a high energy synchrotron, synchrocyclotron or isochronous cyclotron (1-100 GeV).

Future (ionic) engine will have mandatory a circular particle accelerator (high or very high energy; see the Figure 3).

Sure that the difficulties will arise from design, but they need to be resolved step by step.

We can thus increase the speed and autonomy of the ship using a less quantity of fuel.

One can use synchrotron radiation (synchrotron light, high intensity beams), like high intensity (X-ray or Gamma ray) radiation, as well. In this case will be a beam engine (not an ionic engine).

A linear particle accelerator (also called a LINAC) is an electrical device for the acceleration of subatomic particles. This sort of particle accelerator has many applications. It used recently as to an injector into a higher energy synchrotron at a dedicated experimental particle physics laboratory. In this, the big classic synchrotron is reduced to a ring surface (magnetic core).

The design of a LINAC depends on the type of particle that is being accelerated: electron, proton or ion.

It proposes using a powerful LINAC at the exit of synchrotron (especially when one accelerates electrons) to not lose energy by photons premature emission (figure 3).

One can use a LINAC in the entry in the Synchrotron and one at out (Figure 2). To use a small entrance LINAC, between him and synchrotron, one put an additional speed circuit in a stadium form (Fig. 2).

The end LINAC can be reduced if one put more end LINACs. See diagram below (Fig. 2.).

Fig. 2: *A high energy synchrotron schema*

This ship has two circular particle accelerators (two synchrotrons)

This ship has first a circular particle accelerator (a synchrotron), and at the end two big linear particle accelerators (two big LINAC)

Fig. 3: *Some flying synchrotron prototypes*

CONCLUSIONS

Speaking about a new ionic engine means to speak about a new aircraft.

The chapter presents shortly the actual ionic engines (called ion thrusters) and the new ionic (pulse) engines proposed by the author. Ionic engine (ion thruster, which accelerates the positive ions through a potential difference) is about 10 times more effective than classic system based on combustion.

We can still improve the efficiency of 10-50 times if one uses pulses of positive ions accelerated in a cyclotron mounted on the ship; the efficiency can easily grow for 1000 times if the positive ions will be accelerated in a high energy synchrotron, synchrocyclotron or isochronous cyclotron (1-100 GeV).

Future (ionic) engine will have mandatory a circular particle accelerator (high or very high energy). We can thus increase the speed and autonomy of the ship using a less quantity of fuel and power. One can use synchrotron radiation (synchrotron light, high intensity beams), like high intensity (X-ray or Gamma ray) radiation, as well. In this case will be a beam engine (not an ionic engine), it'll use only the power (energy, which can be solar energy, nuclear energy, or both) and so we will remove the fuel.

A linear particle accelerator (also called a LINAC) is an electrical device for the acceleration of subatomic particles. This sort of particle accelerator has many applications. It used recently as to an injector into a higher energy synchrotron at a dedicated experimental particle physics laboratory. In this, the big classic synchrotron is reduced to a ring surface (magnetic core).

The design of a LINAC depends on the type of particle that is being accelerated: electron, proton or ion.

It proposes using a powerful LINAC at the exit of synchrotron (especially when one accelerates electrons) to not lose energy by photons premature emission (figure 3).

One can use a LINAC in the entry in the Synchrotron and one at out (figure 2). To use a small entrance LINAC, between him and synchrotron, one put an additional speed circuit in a stadium form (fig. 2).

With a new ionic engine one builds a new aircraft, which can travel through water and. This new aircraft will can accelerate directly, without an additional combustion engine and without gravity assists from other planets

Ionic engine (ion thruster) has 2 major advantages (a) and 2 disadvantages (b) compared with chemical propulsion; (a) the impulse and energy per unit of fuel used are much higher; 1-the increased impulse generates a higher speed (velocity; so we can walk longer distances in a short time), 2-the high energy decreases fuel consumption and increase the autonomy of the ship; (b) generate force and acceleration are very small; we can't defeat any forces of resistance to lodging by atmosphere and we have no chance to exceed gravitational forces - ship will not leave a planet (or fall on it) using the ion thruster (It required an additional motor). Vacuum ship acceleration is possible but only with very small acceleration.

Increasing more the energy (and also the impulse) can reach the necessary forces and acceleration (Growth will need to be very high, 100 PeV-1000 PeV). Particles energy increased can be made with accelerators circular and or modern linear. Particles energy increased will be huge and in addition will need to grow and the flow of accelerated particles (and the tor diameter; if one increases enough the flow, the necessary energy will be 10 GeV-10 TeV).

Immediate consequence of increasing particle energy will be the increasing of speeds and autonomy of the ship. Now we can achieve huge speeds in a very short time. The ship will pass through any atmosphere (including water) with great ease. The ship can take off or land directly.

Initially one can use to ship the old forms (the old design) which adapts and the accelerator(s).

REFERENCES

[1] Wikipedia, *the free encyclopedia*, net,

[2] Dan Tanna, *Technology today*, edit on 10-6-2008, a net Link.

CHAPTER V - CAPTURING ENERGY CONCENTRATED NEAR THE SOURCE AND FORWARDING DIRECTLY TO EARTH IN CONCENTRATED FORM

CAPTURING ENERGY CONCENTRATED NEAR THE SUN

Should start some spatial projects, to capture a large amount of energy somewhere near the source (near the Sun), energy which can be sent then to the Earth in a concentrated form (LASER, MASER, IRASER, etc).

The enormous energy emanating from the sun is spreading in all directions of the universe, and dilute with the distance.

On Earth no longer reach than a small amount from the energy emanated by the sun.

We try here (on the Earth) to capture a drop from a very small amount of energy, who came from Sun. And we also complain that the yield is low, and technological costs are high.

A large amount of energy is transmitted to long distances with low losses, naturally, because is emitted by a sun (a star) in concentrated form, with natural radiations.

Eta Carinae is a stellar system in the constellation Carina, about 7,500 to 8,000 light-years from the Sun. The system contains at least two stars, one of which is a Luminous Blue Variable (LBV), which during the early stages of its life had a mass of around 150 solar masses, of which it has lost at least 30 since. It is thought that a Wolf-Rayet star of approximately 30 solar masses exists in orbit around its larger companion star, although an enormous thick red nebula surrounding Eta Carinae makes it impossible to see optically. Its combined luminosity is about four million times that of the Sun and has an estimated system mass in excess of 100 solar masses.

This is exactly what should we do. This sun strange and extremely rare in Universe, shows us what must we do.

The third halo of our sun surrounds the planets Mercury and Venus, and barely touching the Earth.

The fourth halo (the most pale from those which are visible with the naked eye) reach Jupiter.

Mercury is hot, and Saturn is cold.

Installations which must do capturing the solar energy, could be installed over the Mercury.

From the Mercury, the concentrated energy will be transmitted directly focused on the Moon.

On the Moon, the energy will be conserved and forwarded to Earth in doses non-hazardous (with lower

concentrations), using multi-channels microwaves.

SEE YOU SOON!

www.ingramcontent.com/pod-product-compliance
Lightning Source LLC
Chambersburg PA
CBHW071622170526
45166CB00003B/1152